Bibliografische Information der Deutschen Nationalbibliothek:

Die Deutsche Bibliothek verzeichnet diese Publikation in der Deutschen Nationalbibliografie; detaillierte bibliografische Daten sind im Internet über http://dnb.d-nb.de/ abrufbar.

Impressum:

Druck und Bindung: Books on Demand GmbH, Norderstedt Germany
ISBN: 9783346111081

Dieses Buch bei GRIN:

https://www.grin.com/document/512859

Gezahagn Kudama

Impacts of Climate Change on Coffee Production and Possible Measures in Ethiopia

GRIN Verlag

A Seminar Paper

on

Impacts Climate Change on Coffee Production and Possible Measures in Ethiopia

Paper presented at

Fuculty of Resource Economics and Management, Wollega University, Shambu Campus
Agricultural Economics Department

By

Gezahagn Kudhama (Msc)

Lecturer at Wollega University, Shambu Campus P.O. Box 138 Shambu, Ethiopia

December 2019
Shambu, Ethiopia

Abstract

Climate change refers to a change in average weather conditions or the variation of weather within the context of longer-term average conditions. This paper reviews that the impacts of climate change on coffee production and potential adaptation measures in coffee sub-sector in Ethiopia. Accordingly, Ethiopian coffee industry is potentially at risk due to the negative impact of climate change on suitability of coffee production. Climate change been severely affecting the national income of the country in general and the livelihoods of poor smallholders in particular via undermining coffee yield and quality. Furthermore, several studies predicted that will continue to impact coffee production in the future. Accordingly, a policy measure that enforces and promotes irrigation, mulching, pruning, tree shade management, drought, pests and diseases resistant high yield improved coffee variety development for coffee production is paramount

Chapter One
Introduction

1.1. Background

Since Arabica coffee (*Coffea arabica L.*) distributed from Ethiopia and spread to the Yemen peninsula, possibly as early as 575 CE (Anthony, et al., 2002), coffee has become an important agricultural commodity worldwide with more than a billion cups consumed every day. In 2014, more than 8.5 million tons were produced by 26 million coffee farmers in 52 countries with an export value of 39.3 billion US$ (ICO, 2016) as cited in (Hironsa, et al., 2018). Ethiopia is leading Arabica coffee producer and exporter in Africa (USDA, 2018). Coffee is the mainstay of the economy of Ethiopia via contributing 34% of total national export earnings in 2017/18 (USDA, 2019) and coffee farming is income basis for millions of poor smallholders as it offers living revenue for estimated 15 million Ethiopians, 15% of the population (Tefera & Tefera, 2015).

Climate refers to a long term average weather situation of a region that is measured by evaluating the patterns of deviation in temperature, humidity, air pressure, wind, precipitation, evaporation and other environmental variables of that particular region over long periods of time. Any imbalance of environment is happening through instances of heavy disturbances of climate that is known climate change. In other word, climate change refers to a change in average weather conditions or the variation of weather within the context of longer-term average conditions (Van den Bossche, 2017).

Scientists around the world now agree that the reasons of climatic changes occurring internationally are the result of human activity. Even though responsibilities for the origins of climate change primarily with the industrialized nations, the costs of climate change will be borne most directly by the poor. Each and every component of environment is affected by climate change (Scavia, et al., 2002).

Climate change already had a severe and negative effect on coffee production on coffee production throughout the world and it will also continue to impact in the future. As in another place, coffee in Ethiopia is vulnerable to climate change. Modelling studies by Moat, et al. (2017) indicates that the area of bio-climatically suitable space of Arabica coffee could decline approximately between 39 and 59% by the end of the century, based on the emissions scenario. In Ethiopia, rapidly increasing temperatures kill the plants at an alarming rate (Joel, 2014). Pests and disease that target coffee plants have risen in prevalence. Unless action is taken to slow down climate change or find alternative ways to keep coffee plants alive, Ethiopia will be in danger due to the importance of coffee production for their economies (Joel, 2014).

Growing recognition of the vulnerability of coffee to climate change has bigger interest in developing adaptation and coping strategies in the sector in Ethiopia (Kew, 2013) and the ability communities to cope with external stresses is viewed as an imperative in Ethiopia's coffee sector, enabling it to continue to contribute to the long-term economic and social well-being of the country as well as coffee companies and millions of consumers (Hironsa, et al., 2018).

1.2. Objectives of the seminar

The general objective of this seminar is to present the impacts of climate change on Arabica coffee production and potential adaptation measures to be taken so as to withstand with adverse effects of climate change in Ethiopia.

The specific objectives of the seminar are:

1. To review the impacts of climate change on coffee production
2. To assessment potential adaptation options

Chapter Two
Materials and Methods

This seminar paper is completely a review paper and relies on the secondary data. Accordingly, different published journals and reports largely supported in providing data in this paper. Therefore, no specific method has been followed in preparing this paper. The paper has been prepared by browsing internet, studying comprehensively various articles published in different journals, books, proceedings, dissertation obtained from different sources. All the information collected from the secondary sources have been compiled systematically and chronologically to enrich this paper.

Chapter Three
Review of Findings and Discussion

3.1. Overview of Coffee Sector in Ethiopia

Ethiopia can be considered as the biological and cultural home of coffee as well the main storehouse of genetic diversity for Arabica coffee. Ethiopia is Africa's biggest coffee producer and the world's fifth largest exporter of Arabica coffee (International Coffee Organization (ICO), 2015). Even though productivities (kg/hectare) are low compared to other producing countries coffee farming alone provides a livelihood income for around 15 million people (Tefera & Tefera, 2015) and it is single most important source of income for many of coffee producers in the country (Environment and Coffee Forest Forum (ECFF), 2017).

Ethiopian coffee is largely grown under the shade of trees (shade or forest coffee), either within forest or forest-like environments, or in farming systems that incorporate specific shade plants – typically indigenous trees, or sometimes fruit trees and other crop plants. In some areas coffee is grown with little or no shade (sun coffee). Forest (shade) coffee and sun coffee can be regarded as the two main coffee production systems in Ethiopia (Environment and Coffee Forest Forum (ECFF), 2017).

Arabica coffee grows fruitfully within a narrow range of environmental conditions. Coffee will become increasingly stressed as the air temperature rises and soil moisture declines (due to lack of rainfall), and vice versa. Ethiopian coffee can habitually be considered as organic by default, and may possibly exceed the standards set for organic certification. Additionally, coffee production system of the country offers protection for biodiversity and other environmental benefits (Environment and Coffee Forest Forum (ECFF), 2017).

3.2. Ecological Requirements of Coffee Arabica

For optimum growth Arabica coffee desires about 15 to 25 degree Celsius, with well distributed rainfall, ranging from 1,600 to more than 2,000 mm, for a dry season lasting three to four months coinciding with the coolest period and at altitudes ranging from 1,500 to 2,800 meter above sea level (Davis, Gole, Baena, & Moat, 2012).

Whilst erratic rainfall at the beginning of the rainy season can cause abortion of flowers, high temperatures (> 28-30°C) have been observed to reduce flower bud formation and cherry production (Davis, Gole, Baena, & Moat, 2012). On the other hand, increasing temperatures create a more favorable environment for the Coffee Berry Borer and can cause to increase losses by insects. Furthermore, droughts are expected to cause production losses and, in extreme cases, death of coffee trees (Moat, et al., 2017).

3.3. Observed and Predicted Effects of Climate Change in Coffee Producing Areas in Ethiopia

In Ethiopia from 1960 to 2006 mean annual temperature increased by 1.3°C (USAID, 2016). Other observations also indicates significantly increasing trends in the frequency of hot days, and much larger increasing trends in the frequency of hot nights; the frequency of cold days has decreased significantly in all seasons except the dry season (December, January and February) for the same period (McSweeney, New, & Lizcano, 2010). The average annual temperature is projected to increase 1.1-3.1°C by 2060, with hot days and nights expected to become more frequent (USAID, 2016).

It has been stated that the strong variability within Ethiopia's annual and decadal rainfall makes it difficult to detect long term trends, and that there is no statistically significant trend in observed mean rainfall in any season for which climate data is available (1960–2006) (McSweeney, New, & Lizcano, 2010). However, based on data from quality-controlled climate station observations, it has been shown that spring and summer rains have declined by 15–20% since the mid-1970s and late 2000s, in southern, south-western and south-eastern Ethiopia (Funk, et al., 2008). Similarly, 10% reduction of rainfall of around across Ethiopia has been reported (1948–2006) (Jury & Funk, 2013).

Projections from different General Circulation Models (GCMs) are generally consistent in indicating increases in annual rainfall in Ethiopia (Niang, et al., 2014). Overall, projections based on GCMs for Ethiopia are highly variable across regions (Conway & Schipper, 2011) and for some regions the various GCMs do not agree on the direction of precipitation trend.

3.4. Consequences of Climate Change to Coffee Production in Ethiopia

Some coffee growing areas in Ethiopia are already poorly suited for growing coffee, due to the negative of climate and will continue to be impacted so in the future (Environment and Coffee Forest Forum (ECFF), 2017).

Environmental stresses can influence coffee growing in several of ways, from slight physiological stress to the obvious physical manifestations of water and/or heat stress, from wilted leaves (mild stress) to the death of the plant (severe stress). Coffee plants may also show climate-related stress through the increased occurrence, and increased intensity (and diversity), of diseases and pest attacks, which in turns leads to decline in production. During extended drought episodes, for example, there is often a reduction in coffee bean size, and an increase in bean malformations, which are often referred to as 'faults'. Faults, can refers to black marks, blackened beans, misshapen beans and insect damage, are measured by industry standards and have a direct influence on coffee prices, in blend with cup quality. Cup quality (how good or bad the coffee smells and tastes) is assessed by rigorous, standard protocols, in combination with the number of faults. These quality scores place coffee into a category, which traders use to set the grade for the coffee and therefore its associate with price. Even somewhat small changes in the coffee growing climate, within each year's growing season (like low rainfall), can influence yield and cup quality. Such associations between lower quality coffee and the environment have been observed in several coffee growing areas in Ethiopia: like Zege/ Zeghie Peninsula (the northern part of the Amhara coffee area), Harar, eastern Wellega, Bench Maji, and the Rift areas (Environment and Coffee Forest Forum (ECFF), 2017).

Davis, Gole, Baena, & Moat (2012) forecast an 85% reduction of localities where indigenous coffee Arabica is growing naturally by 2080 in Ethiopia. Similar projection indicates that areas becoming suitable for Arabica coffee production are located at higher elevations, previously too cool for Arabica. Cultivation of coffee is starting in some of these areas (USDA, 2018). However, relocating Arabica coffee plants in Ethiopia is not necessarily a good idea because it will take many years for the plants to become productive again (Davis, Gole, Baena, & Moat, 2012).

3.5. Suitable Adaptation Measures

Irrigation: The use of additional water to boost and/or maintain soil water availability for coffee plants is an earliest and highly effective means of improving yield and quality. Arabica coffee can be successfully grown in places that would otherwise be unsuitable due to low rainfall under good irrigation practices. There are many examples in Ethiopia, where good to excellent coffee is grown with low expense irrigation systems (Environment and Coffee Forest Forum (ECFF), 2017).

Tree shade management: Effective shade management can improve coffee growing conditions by reducing daytime air and soil temperatures, and increasing humidity and soil moisture (Jaramillo, et al., 2011). The difference in daytime temperatures between shaded and unshaded coffee plots can be about 4°C. Choice of shade tree species, planting density and canopy coverage are key factors in good shade management for reducing temperature and the negative effects of heavy rain (Environment and Coffee Forest Forum (ECFF), 2017).

Mulching and Pruning: Covering the soil with different materials (e.g. compost, manure) not only helps to preserve soil moisture and decrease soil temperature (reducing evapotranspiration), but it can also increase soil fertility, suppress weeds, and improve rainfall penetration into the ground. Pruning can also help to improve water-use efficiency in coffee plants (Environment and Coffee Forest Forum (ECFF), 2017).

Furthermore, drought, pests and diseases resistant high yield improved coffee variety development can play a crucial role to withstand adverse effects of climate change.

Chapter Four
Conclusion

Despite coffee plays principal role in offering living revenue for many poor smallholders and a backbone of national economy of Ethiopia, the country's coffee industry is potentially at risk due to the negative impact of climate change on suitability of coffee production in the country. It has been severely affecting the national income of the country in general and the livelihoods of poor smallholders in particular via undermining coffee yield and quality. Furthermore, several studies predicted that will continue to impact coffee production in the future. Accordingly, a policy measure that enforces and promotes irrigation, mulching, pruning, tree shade management, drought, pests and diseases resistant high yield improved coffee variety development for coffee production is paramount.

Bibliography

Anthony, F., Combes, M., Astorga, C., Bertrand, B., Graziosi, G., & Lashermes, P. (2002). The origin of cultivated Coffea arabica L. varieties revealed by AFLP and SSR markers. *Theor. Appl. Genet.* , (104): 894–900.

Conway, D., & Schipper, E. L. (2011). Adaptation to climate change in Africa: challenges and opportunities dentified from Ethiopia. *Global Environmental Change*, (21): 227–237.

Davis, A. P., Gole, T. W., Baena, S., & Moat, J. (2012). The Impact of Climate Change on Indigenous Arabica Coffee (Coffea Arabica): Predicting Future Trends and Identifying Priorities. *Plos One*, 7(11): 1-14. https://doi.org/10.1371/journal.pone.0047981.

Environment and Coffee Forest Forum (ECFF). (2017). *Coffee Farming and Climate Change in Ethiopia Impacts, Forecasts, Resilience and Opportunities.* . Environment and Coffee Forest Forum. Summary Report 2017.

Funk, C., Dettinger, M. D., Michaelsen, J. C., Verdin, J. P., Brown, M. E., Barlow, M., & Hoell, A. (2008). Warming of the Indian Ocean threatens eastern and southern African food security but could be mitigated by agricultural development. *Proceedings of the National Academy of Sciences of the United States of America*, (pp. (105): 11081–11086).

Hironsa, M., Mehrabib, Z., Gonfac, T., Morela, A., Golec, T. W., McDermotta, C., . . . Norris, K. (2018). Pursuing climate resilient coffee in Ethiopia – A critical review. *Geoforum*, (91): 108–116.

International Coffee Organization (ICO). (2015). *Historical Data on the Global Coffee Trade.* International Coffee Organization (ICO). Website: http://www.ico.org/new_historical.asp?section=Statistics.

Jaramillo, J., Muchugu, E., Vega, F. E., Davis, A., Borgemeister, C., & Chabi-Olaye, A. (2011). Some Like It Hot: The Influence and Implications of Climate Change on Coffee Berry Borer (Hypothenemus hampei) and Coffee Production in East Africa. *Plos One*, 6(9): 1-14.

Joel, I. (2014). The Impact of Climate Change on Coffee Production in Colombia and Ethiopia. *Global Majority E-Journal*, 5(1): 33-43.

Jury, M. R., & Funk, C. (2013). Climatic trends over Ethiopia: regional signals and drivers. *International Journal of Climatology*, (33): 1924 - 1935.

Kew, S. (2013). *Building a climate resilient coffee economy for Ethiopia.*

Kifle, B., & Demelash, T. (2015). Climatic Variables and Impact of Coffee Berry Diseases in Ethiopian Coffee Production. *Biology Agri and Health Care*, 5(7): 55-64.

McSweeney, C., New, M., & Lizcano, G. (2010). *UNDP Climate Change Country Profiles: Ethiopia.*

Moat, J., Williams, J., Baena, S., Wilkinson, T., Gole, T. W., Challa, Z. K., & Demissew, S. (2017). *Resilience potential of the Ethiopian coffee sector under climate.*

Niang, I., Ruppel, O. C., Abdrabo, M. A., Essel, A., Lennard, C., Padgham, J., & Urquhart, P. (2014). *Africa. In: Climate Change 2014: Impacts, Adaptation, and Vulnerability Part B: Regional Aspects. Contribution of Working Group II to the Fifth Assessment Report of the Intergovernmental Panel on Climate Change.* New York: Cambridge University Press.

Scavia, D., Field, J. C., Boesch, D. F., Buddemeier, R. W., Burkett, V., Cayan, D. R., & Reed, D. J. (2002). Climate change impacts on US coastal and marine ecosystems. *Estuaries*, 25(2):149-164.

Tefera, A., & Tefera, T. (2015). *Ethiopia: Coffee Annual Report (Global Agricultural Information Network, GAIN Report Number: ET- 1302).* Addi Ababa.

USAID. (2016). *Climate Change Risk Profile: Ethiopia.* USAID. Retrieved from https://www.climatelinks.org/sites/default/files/asset/document/2016 CRM.

USDA. (2018). *Ethiopia Coffee Annual (Global Agricultural Information Network, GAIN Report Number: ET1820).* Addis Ababa: United State Department of Agriculture Foreign Agriculture Service.

USDA. (2019). *Ethiopia Coffee Annual Report (Global Agricultural Information Network, GAIN Report Number: ET1904).* Addis Ababa: United State Department of Agriculture Foreign Agriculture Service (USDA).

Van den Bossche, A. (2017). Global warming and sea level rise: faster than expected. *Global warming and sea level rise: faster than expected*, (pp. 1-27).